wild
puppies

This book is dedicated to all the organizations and individuals who work each
day for the benefit of the wild nondomestic dogs of the world. These include
the Canadian Nature Federation, Defenders of Wildlife, National Audubon Society,
The National Wildlife Federation, The Sierra Club, The Nature Conservancy, The
Wilderness Society, The World Wildlife Fund, and all the less known but equally
valuable groups around the world from which most of us have yet to hear.

wild puppies

by peggy bauer, photographs by peggy and erwin bauer

CHRONICLE BOOKS

SAN FRANCISCO

Book and cover design by Nancy Zeches, San Francisco. Printed in Hong Kong.

Library of Congress Cataloging-in-Publication Data available. ISBN: 0-8118-1039-9

Distributed in Canada by Raincoast Books, 8680 Cambie Street, Vancouver, B.C. V6P 6M9. 10 9 8 7 6 5 4 3 2 1

Chronicle Books, 275 Fifth Street, San Francisco, CA 94103

Photographs: front cover, Gray wolf pup; back cover, Silver-backed jackel pup; page 1, Silver-backed jackal pup.

The maps in this book indicate approximate ranges for the various species.

contents

When the pups feel threatened, they huddle together for mutual comfort. They might also scramble into a handy hole, perhaps an abandoned badger or fox den.

The dog is one of the most familiar animals to humans. It has been estimated that 38 percent of American households and 28 percent of Canadian households have at least one dog. Dogs are guides for the blind, guards for flocks of sheep, hunting allies, and faithful companions.

The dog's wilderness cousin, the wild dog, is not, however, as beloved. In fact, wild dogs are often in conflict with humans and compete, if not for the same food, then for the same territory. Wild dogs live on virtually every continent, from the Arctic Circle to the tip of South America and the Cape of Good Hope, although they are believed not to be indigenous to Antarctica, New Zealand, Madagascar, or any of the Pacific islands.

Their wide geographic distribution is matched by their diversity. The largest, the gray wolf, may weigh 100 pounds; the smallest, the fennec fox, often weighs less

introduction

than two pounds. The South American bush dog can dive underwater; the gray fox climbs trees; Arctic foxes thrive in frigid temperatures; and fennec foxes survive the Sahara's blistering heat.

Wild dogs have few natural enemies, yet many species are endangered, primarily because of humans. The destruction of habitat and organized campaigns to eliminate certain species have meant that at least half of the thirty-five species of wild dogs are threatened with extinction.

Comparisons between wild dogs and wild cats are inevitable. The dogs are more social. They cooperate in hunting and raising their young—male cats leave the females after mating. Wild dogs have a large variety of prey, whereas cats are more limited in their choices. Although some cats are very swift to strike, they are generally stalkers, attacking from ambush, whereas certain dogs, such as wolves and African wild dogs, are adapted to chases over miles of ground. Some cats can be many times larger than the dogs. This has been attributed to their solitary hunting style, which demands power and strength to succeed.

Wild dogs generally have more offspring than do cats, but each pup is smaller compared to its eventual adult weight and requires more attention from family members than young cats. Wild dogs—either a pack or a family group—care for the pups for several weeks or months, bringing them food until they can hunt for themselves. Young cats seem to adapt more quickly to hunting.

In the range and variety of their habits, wild dogs are a wonderful study. They are social and they are loners, predators and prey. The study of their often complex lives can contribute much to our sense of environmental forces and to our understanding of our place within the natural world.

Largest of the wild dogs, wolves live in packs of five to fifteen animals that are organized around a dominant (alpha) male and female, which are the only breeding pair in the group. The pack has a complex social organization, with strict hierarchies and rules for both sexes. This arrangement permits cooperative hunting of large prey such as deer and elk. Because wolves are extremely adaptable, they also catch beaver and other small animals if given the chance. In the summer, certain members of the pack may leave for extended periods of time, but they inevitably return to the group.

In Alaska, native Nunamiut Eskimos have traditionally competed with the wolf for food and are intimately acquainted with its skills. They admire it for its intelligence and social organization. An old Siberian proverb states that a wolf is kept fed by its feet. Blessed with great endurance, wolves can lope for hours and are the best runners in the dog world. One pack was tracked as it traveled thirty-one miles per day for nine days. The range of loner wolves without a pack affiliation may be greater than a pack's, since single wolves must avoid the territory of other wolves, which are among the deadliest enemies of a single wolf.

Mating occurs between January and April, and the litter averages six pups. A crevice or ground hole provides space for the den, which may be used for several years. Early in their lives, the pups establish a hierarchy of dominance and subservience that will remain constant even as they mature.

Although heavily hunted in years past, the wolf continues to endure, sustained even in hard times by its cooperative social structure and clever adaptation to changing circumstances.

gray wolf

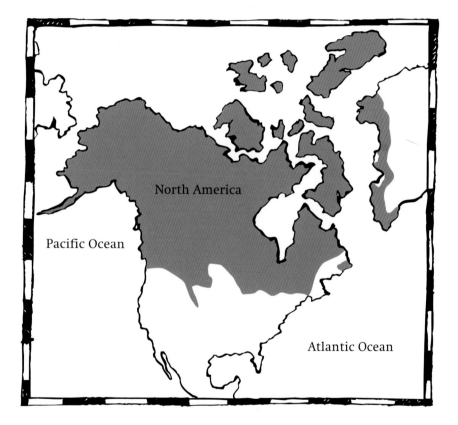

Pacific Ocean

North America

Atlantic Ocean

If there has been some disturbance that upsets the pack, the entire group moves to a new location, often a long distance from the original den site. The young are normally carried carefully, lifted by the loose fold of skin at the back of the neck. They are moved one at a time until all are secure in the new surroundings. Occasionally, an inexperienced mother might grasp the puppy by another part of its body, but no harm seems to be done.

Until the pups are old enough to accompany the adults on hunting trips, they remain behind with a "pup-sitter," which may or may not be the mother.

This puppy is three weeks old. His eyes are still blue, but they will turn gold as he reaches maturity.

One youngster, interested in watching its mother's vigorous digging, wandered into a spray of flying debris and ended up with a face covered with dirt.

The haunting howl of a wolf ranks with the call of a loon or the bugle of an elk in the fall as one of the most wonderful sounds of nature. Howling is thought to serve several purposes: warning off other packs of wolves, keeping in touch with absent pack members, or providing some simple enjoyment. Young pups learn to do it early, usually stimulated by hearing other wolves howl, but even the distant sound of a siren or train whistle can initiate a wolf chorus, as all join in for a great, wild symphony of sound.

By ten weeks of age, the fat, round puppy look is gone, replaced by a gangly adolescent appearance. Now legs lengthen quickly, and the baby hair, which was furry, is replaced by the coarser pelage of an adult wolf.

A hollow log is a good hiding place for a frightened and confused wolf pup. From time to time, the log may already be occupied by a sleeping opossum, a resting hare, or even a snake, but ordinarily such a place provides good security and may be used over and over again as the pups become bolder.

As they grow, the pups play with one another, building strength and coordination. At the same time, they establish a ranking within the group that will not change as they mature. Since only one pair in any pack of wolves will mate and produce young, the most precocious in any litter are the most likely to achieve this position.

Finding such things as a turtle, grasshopper, or old mule-deer skull, as shown here, can hold a pup's attention for some time. The curious pups use all their senses to help them understand the world in which they live.

When the hunting party returns, the young are fed regurgitated food. To receive a meal, the pups must "beg" by whining and gently biting the muzzle of an adult.

After six months, the pups look like small
adults, even like domestic dogs. The cute stage
of life is gone forever.

The gray fox is one of the most adaptable of the wild dogs. It is found in much of the United States as well as Central America and northwest South America. Although moderately abundant, the gray fox is less familiar than the red fox because it is nocturnal, whereas the red fox often hunts boldly in full daylight. Food ranges from plants to carrion, rodents to fruit. The gray fox generally stalks its prey, trying to get as close as possible before making a quick rush for the capture.

Breeding season of the gray fox is determined by climate. Foxes in warm areas breed earlier in the year than their more northern cousins. The males accompanying a female fight vigorously at this time for dominance, and then the female makes a selection of mate. Regardless of the time of breeding, a two-month gestation results in a litter generally consisting of four pups, which are born with their eyes shut and suckle for about ten weeks. The young are extremely quarrelsome, apparently to determine dominance.

The male fox may help with the pups, and the family remains together until the following January or February, when it breaks up, the young males departing first. The young females may remain with the mother until fall. To avoid conflict with existing family groups, the young foxes will travel long distances to establish their own territories. Mortality is extremely high in the first year of life.

Unusual among foxes, the gray foxes are able to climb, so they also feed on birds and berries, and may even make their den in a tree. In addition to hawks that prey on the young, the gray fox's few natural enemies include cougars, bobcats, and coyotes as well as humans.

gray fox

North America

Pacific Ocean

Atlantic Ocean

Dens are used as a place not only to shelter the young, but also to rest and sleep in the winter. Hollow logs on the ground, an old badger hole, or a cozy crevice in a rockpile also serve as den sites. The young gray fox pups here venture out from a den in a hillside to investigate the surroundings and play in the warm sunshine.

Because the gray fox can climb trees, it is also known as a tree fox. The claws on the back legs are longer, sharper, and more curved than those of other foxes, giving the animal an almost catlike ability to scale nearly vertical limbs. The gray fox might use a hollow in a tree trunk some distance above the ground for a den.

The red fox, which is common on four continents, has replaced the wolf as the most widely distributed wild dog. Intelligent and adaptable, red foxes live alone or in hierarchical groups that define their territories based upon the availability of food. Where there is a great variety of prey, the red fox will establish a large territory.

They are active at day and night, and they will eat virtually anything. Reproduction adjusts to the availability of prey, and the red fox also changes the size and composition of its family group in response to changing conditions.

In any group of red foxes, all the females breed, but ordinarily only the dominant vixen bears and raises pups. The mother fox is extremely attentive to the young, often spending hours at play with them. At approximately four weeks, the pups fight furiously among themselves to establish social rank. Sometimes only one pup survives, to live and in its turn to breed superior young. The population is thus controlled, and the best traits are passed on.

Red fox mortality is high. It is hunted, trapped, and poisoned. Other risks include disease and predation by coyotes. But despite their many enemies, red foxes seem far from endangered.

red fox

Pacific Ocean

North America

Atlantic Ocean

Like all young, fox pups play with each other, building strength and coordination. This play frequently degenerates into naked aggression, establishing the dominance of the stronger, more intelligent pup.

Their coats are not yet red, the color they will acquire as they mature. They blend well with the den, and this camouflage protects them from threatening enemies.

The den may be one used for several years (one den was known to have been used by successive generations for thirty-five years). The pups, usually four or five, although up to a dozen have been counted, are born after about fifty days of gestation. A burrow may be occupied by a single male (dog) fox and two females, each having her own brood. The young weigh only about one-quarter pound. The mother stays with the very young pups continually for three or four days, then leaves briefly on hunting forays. Gradually, the youngsters venture out from the den.

In North Dakota, not fifty yards from a busy interstate highway, four small fox pups sit near an almost invisible den. Huge, noisy trucks rumble past the den, raising clouds of dust in their wake. It seems an odd place for a home, but demonstrates the adaptability of the red fox.

By eight weeks, the pups are maturing quickly, showing their natural agility and speed. They can run swiftly, even over uneven ground, and by ten weeks the pups are weaned. Two or three months later, the family disbands.

There are several color variations among red foxes, including cross foxes, which make up about one-quarter of all red fox numbers. Crosses (brown to rust color with a darker swath across the shoulders and down the spine), silver foxes (about 10 percent of all foxes), and red-colored may be found in the same litter.

These cross foxes energetically dig several holes for no apparent purpose, dirt flying out behind as if blown by a huge fan.

Besides playing and fighting, the pups are also entertained by objects such as a passing beetle, a butterfly, or a dead raven. The parent fox has brought the raven to the den, but its large size (nearly the same size as the pups themselves) deters the youngsters from considering it a meal. Each time the pups approach the raven a vagrant breeze flaps a wing menacingly, and the pups scamper away in terror.

These four-week-old youngsters scramble out of sight when surprised at the entrance to their den. But foxes are curious, and one peeped above the ground, soon to be joined by three siblings.

The mother red fox joins in the play, and at dusk the entire group is yapping and growling, pouncing and wrestling, a wildly mobile con-glomeration of mouths, feet, and white-tipped tails tumbling in the last rays of the sun.

The little swift fox, only about the size of a domestic cat, was once abundant on the tall grass prairies that stretched through the middle of North America. When the grassland was plowed for farms, the populations of mice, gophers, and rabbits that were the swift fox's prey decreased and, with them, the swift fox. The fox was also intensively trapped during the nineteenth century and were killed by carcasses laced with poison intended for wolves and coyotes.

Swift foxes are diggers. They generally take over an unused badger or marmot den and create elaborate burrows where they spend much of their time before venturing forth to forage. In fact, the swift fox spends more time underground than any other member of the dog family.

They have no known social organization outside the family group. Swift foxes live in pairs and normally have three to six pups, which are born in the spring. After the arrival of the pups, the father fox has duties at the den from feeding the new born with regurgitated food to defending the den. Pups may be moved over a dozen times in a year, and by August the litter is broken up. Since swift foxes can breed in their first year, they have an excellent chance to increase their numbers, but they compete with coyotes and are the coyote's prey, so their progress in expanding their territories is slow.

swift fox

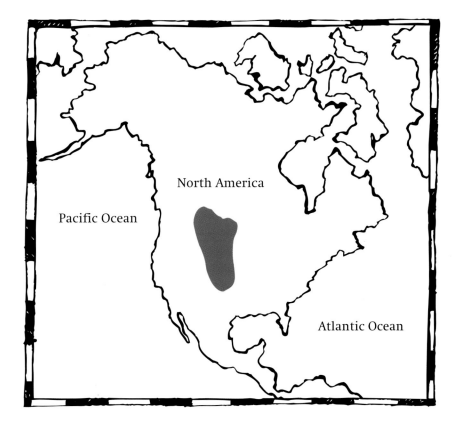

Pacific Ocean

North America

Atlantic Ocean

Swift fox pups have typical canine faces with erect, pointy ears and alert eyes.

A captive swift fox family in southern Alberta, Canada, with pups several months old, overcome the natural fear of humans when the curious pups pop up one by one. Soon they romp as the young do, chasing, ambushing, and pouncing on each other.

These plucky foxes live in a circular region surrounding the North Pole and in coastal areas of Iceland, Greenland, and Siberia. The Arctic fox's range has been decreasing because of increased predation by red foxes. Other enemies include bears and wolves; disease is also a frequent killer. Even at best, survival is tough: Eighty percent of the fox pups that leave the den will not live through the first winter.

To live in such inhospitable and frigid surroundings, the fox has short, round ears and a small muzzle that are resistant to freezing. The abundant fur on its feet provides good insulation and traction in snow. The fox also has a very thick coat with great warming qualities.

The Arctic fox is generally monogamous, the pairs remaining together through the breeding season and longer. Some observers believe these foxes mate for life. They may use the same den for many successive years and may place their den in close proximity to several other fox dens. They breed in March or April, and the pups are born in May or June. A normal litter is five pups. Once the young are born, the father helps to provide food for the newborn pups.

In some parts of the Arctic, these foxes prey on lemmings. In years of high lemming counts, fox litters are very large and nearly all the pups survive. Not surprisingly, a decline in lemmings leads to an inevitable drop in the number of surviving fox pups. Like all foxes, they are adaptable, and also eat fish, carrion, berries, and plants. Although it is small, this fox may travel several hundred miles a year as it migrates across its snowy habitat.

arctic fox

Arctic Ocean

Greenland

North America

Atlantic Ocean

This Arctic fox's den is on a slight slope. The same den is often used by generations of foxes, some for a century or more. As early as March or April, the female fox returns from the sea ice and goes inland to set up her breeding territory. In May or June, after a gestation period of about fifty days, the pups are born. Usually five are in a litter, but the number varies enormously with the available food supply. At birth their eyes are closed and the pups have a dark fuzzy coat and weigh only about two ounces. Both parents share the responsibility of feeding the young. The father is especially helpful, bringing food to the mother and pups for three weeks. Occasionally, an unmated female also helps with the pups. She may be a daughter from a previous litter and remains with the family for six to eight weeks.

A common murre lies beside the den entrance, probably brought by the vixen, but the pups seem uninterested. The pups and their mother are a steely blue color, typical of the Arctic foxes on the Pribilof Islands. The blue phase is more common than the white in the eastern Arctic and coastal Greenland, but white dominates on the Arctic coast of Alaska.

The mother Arctic fox's insistent barking warns the young of danger, but the curious pups soon peep from their den. The more adventurous come into full view in spite of her repeated calls.

The pups first leave the den at one month of age. In Alaska the pups are independent in August and can fend for themselves. By September, all pups are on their own.

In summer seabird colonies are favorite hunting areas for the foxes. They consume their fill of unguarded chicks and eggs and bury those they cannot eat. Often another fox digs up a cached meal, carries it away, and buries it again. Arctic foxes have been observed trotting along behind polar bears, alert for any scraps that might be left from the bear's dinner. The foxes are not shy and have been reported to follow humans and bark at them.

By September the land has frozen and some sea ice has formed. Rather than making the snow difficult for the young, this may increase their feeding opportunities. The foxes spend the winter on ice floes far from land, eating dead seals, walrus, and whales, or, as in summer, following the polar bears to feed on leftovers.

Before the arrival of Europeans in North America, coyotes lived in the southwestern deserts where bison, antelope, and elk roamed in great herds. Wolves followed the herds and kept the coyotes from expanding their range. With the destruction of the wolf by humans, coyotes thrived, spreading north and east across the continent, inhabiting land from Alaska to Central America. They have even crossed to Newfoundland on ice.

Also known as the prairie wolf or brush wolf, the coyote is the subject of countless songs, poems, and stories that tell of its cunning. Coyotes normally live in open grassland, deserts, or sparsely wooded areas, where they do not have to compete with wolves or cougars. They generally do not defend a specific territory but seem more concerned with the range and location of prey. Normally solitary, coyotes may also live in pairs, but they appear to have no larger social organization. In selecting a mate, the female chooses from among several suitors, and this bond may continue for many years. Courtship begins in late December, but mating may not occur until March. After two months, a litter of from two to twelve cubs is born. By nine months, the pups reach adult size and may soon leave the family group.

The natural enemies of coyote pups are raptors (hawks and eagles), packs of domestic dogs, other coyotes, and disease. Coyotes eat rodents, rabbits, and hares, but they also consume grasshoppers, berries, carrion, and fruit. They seem to prefer fresh meat but will eat carrion if necessary.

In this century, the opportunistic coyote has greatly extended its range, usually into areas where other wild dogs such as wolves suffered a decline in population. This versatile animal is unlike any other member of the wild dog family. Whereas other wild dogs may cling tenuously to existence, the coyote seems to prosper virtually anywhere, eating anything, reproducing under the most unlikely circumstances, and outwitting many human attempts to control it.

Pacific Ocean

North America

Atlantic Ocean

Coyote pups stay safely in the den for the first two weeks of life. Soon after, their eyes open and they venture into the world outside. From the beginning, some pups are larger, more adventurous, and stronger than their siblings, and this superiority generally continues as they mature.

At three weeks of age, after they are weaned, the littermates emerge from the security of the burrow. At this early stage of their lives, they do not wander far and are ready to retreat to the den at a moment's notice.

Mounting is a dominance mechanism among males, and the practice begins early. Only when the pecking order has been established to everyone's acceptance, if not satisfaction, will the aggression stop.

Older coyote pups wander, the males usually more widely and more often than the females. Their play often turns to fighting that establishes a lasting hierarchy.

Howling seems instinctive with coyotes, as it is with wolves. Long ago it was so familiar to Mexicans that they called the coyote "el soprano." Howling may be done to warn away other coyotes or to maintain communication among scattered members of the same band.

Pouncing on a mouse seems as instinctive in a coyote pup as in any cat, but coyote pups track and pounce on almost any moving object in their immediate vicinity.

Coyote pups are intelligent and can be interested in almost anything close to the den area. Investigation of an old deer carcass takes a lot of time. Not only the sight, but the smell and feel of the skeleton intrigues the young pups.

By eleven weeks the cute baby roundness of
the puppies has given way to the gangly stage
when knees look far too large for the size of
the animal, the ears seem oversized, and the
paws seem to belong to a much larger creature.

At three weeks of age, after they are weaned, the littermates emerge
from the security of the burrow even more frequently, although at this
early stage, they do not wander far and are ready to retreat to the
den's safety. Now the mother emerges too, sometimes with the pups,
and at other times by herself, for a few minutes of peace and solitude.

other wild dog pups

Fennec fox litters are born from March to May in the North African deserts after a gestation period estimated at about fifty days. The young usually number from two to five. If the first litter is lost, the pair produces a second. Young pups' coats have dark patches, but after two weeks they are light beige, and by three months the color is like an adult's. They first venture out of the den at four weeks and continue to nurse for another month.

Little is known of several other wild dogs that are found in small groups throughout the world. Most live in remote and inhospitable locations and are often nocturnal.

African wild dog Also called the Cape hunting dog, the African wild dog lives in large groups that help each other to hunt and raise the young. Ordinarily, only the dominant male and female of the pack produce a litter, which generally numbers ten pups. At three months, the young are able to accompany the adults on hunts, but their hunting skills are not fully developed until they reach one year of age.

These dogs are efficient predators that run their prey to exhaustion before finishing it. The hunting dogs appear to have specialties, one group taking turns in the attack and chase, another performing the final killing phase, and still others remaining with the pups to guard them. In recent years, canine distemper has wiped out a large percentage of these brightly patterned dogs.

Bat-eared fox A small fox of eastern and southern Africa, these foxes are easily identified by their large ears, which help them to pursue their prey and to dissipate heat. Although they eat mainly insects, they are hunted by humans who wrongly believe the foxes attack farm animals.

Black-backed jackal This dark canine is one of two strictly African jackals, the other being the side-striped. The golden jackal, which also lives in Asia, is the third type of jackal in Africa. The black-backed jackal survives in two distinct areas: one group is in southern Ethiopia, Sudan, Somalia, Kenya, and Uganda; the second group covers the area from the Cape of Good Hope to Angola, Zimbabwe, and Mozambique.

Mating takes place in the late spring or summer, followed by a two-month gestation period. The den is usually built in an abandoned termite mound. Litters usually number two to seven, and the cubs remain in the den for about twelve weeks. Both parents share in the raising of the pups, and often a young member of the family also assists with new cubs. Among the new cubs, which fight regularly, dominance may determine which young must leave the den.

Dhole The dhole is similar to a medium-sized, short-haired dog, red or mahogany in color. It once ranged through eastern and central Asia, but its present range is greatly reduced. Nonetheless, it is still sufficiently wide-ranging to be found pursuing sea turtles near the ocean and sheep in Asian mountains.

They breed at any time of the year and have a gestation period of sixty days. A litter contains four to five pups. Several families may share a single den, and the mothers and young are fed by the pack.

Dholes hunt in packs of up to forty animals. They are not swift, but they are determined hunters that cover a wide territory in search of food. They eat deer and similar creatures, but they can also bring down water buffalo. Reportedly, dholes have also attacked bears and tigers.

Dingo Aborigines migrating from New Guinea are thought to have brought the dingo to Australia. At one time the dogs lived throughout the continent, but efforts to protect sheep from dingo attacks have drastically reduced the dog's numbers.

Little is known about the animal's behavior in the wild. Four to five pups are born in late winter or early spring, usually in an abandoned rabbit burrow. Both parents cooperate in caring for the young, which mature after about two years.

The dingo has few enemies except humans. Its attacks on sheep in Australia have led to several serious attempts to eliminate dingo populations. Although these efforts have resulted in greatly reduced numbers, the wild dogs still survive.

Ethiopian wolf This wolf struggles to exist in Ethiopia's national parks. It is threatened not only by local herding but by its breeding with domestic dogs, which reduces its ability to survive in the wild.

Fennec fox Except for their extraordinarily large ears, the fennec foxes look surprisingly like the Arctic fox. They are as well adapted to their desert environment as the Arctic fox to the polar regions. Fennec foxes are only about twelve inches long, with an additional six inch tail. They are light-colored with dense fur, even on the bottoms of their feet, to insulate them from the freezing desert nights and scorching days.

Fennec foxes are the smallest of all wild dogs and the most elusive. They live in communal groups of ten to fifteen. During the day, they remain in their burrows to avoid the heat, emerging at night to search for insects, locusts, and lizards. These wild dogs are hunted intensively in the Sahara for sale as food, but they also make fine pets that soon rid any home of scorpions and snakes.

Iberian wolf Very few Iberian wolves still survive in Portugal and Spain, in large part because domestic sheep are important in the Iberian peninsula and people believe the wolves prey on livestock.

Maned wolf This animal bears only a slight resemblance to other wolves. Its ears are particularly large, and its legs enormously long. The long, dark hairs growing from its neck give the wolf its name. Although other wolves are sociable, the maned wolf is a loner, the least social wild dog. It is also apparently guileless and shy.

Common from the Amazon Basin to Argentina, this wild dog is not a swift runner, despite its long legs. Its jaws are weak and its teeth small. It cannot, therefore, raid sheep herds but feeds on insects, rodents, and a South American fruit resembling a large tomato.

The dens of the bat-eared fox are inconspicuous holes in the ground. In East Africa, animals are not seen in the very early morning, but they can often be found when the sun is high. About midday, the pups and the adults sun themselves near the burrow entrance.

Raccoon dog This remarkable animal is unlike other wild dogs. While it appears to have originated in eastern Asia, the raccoon dog has done extremely well in other areas to which it was relocated, living now even in France and England. This success continues today, except in Japan, where raccoon dogs began and where they are now rare.

The pups are born in early summer in litters of three to eight. Blind at birth, they are covered with soft, dark fur and weigh two to three ounces. Although the pups may mature at six months, the family does not break up. Raccoon dogs are the only wild dogs that hibernate. In the summer, they live in the mountains, moving to the valleys in the winter where they may sleep for long periods of time. Normally, they eat small animals and plants but will pillage garbage cans if given the chance.

Red wolf These wolves once thrived from Florida to central Texas and northward along the Mississippi River. However, they were gradually displaced by the arrival of coyotes that competed with the red wolves for food and even bred with them, to the detriment of the wolf population. Today, the adaptable coyote remains the biggest threat to the red wolves' continuation.

Red wolf pups are usually born in the spring in litters of two to eight. Both parents are present for the young, which remain with the family for two years or longer.